Geography Zone: Landforms™

HILLS

Emma Carlson Berne

PowerKiDS press.

New York

Published in 2008 by The Rosen Publishing Group, Inc.
29 East 21st Street, New York, NY 10010

First Edition

Editor: Joanne Randolph
Book Design: Julio Gil
Photo Researcher: Jessica Gerweck

Photo Credits: Cover © iStockphoto.com/Peeter Viisimaa; p. 5 © iStockphoto.com/David MacFarlane; p. 7 © iStockphoto.com/Steve Geer; p. 9 © iStockphoto.com/Alexander Hafemann; p. 11 © iStockphoto.com/Jennifer Stone; p. 13 © iStockphoto.com/David Stearn; p. 15 © Phil Schermeister/Corbis; p. 17 © iStockphoto.com/Paul Tessier; p. 19 © iStockphoto.com/Bonnie Jacobs; p. 21 © iStockphoto.com/Jonathan Larsen.

Library of Congress Cataloging-in-Publication Data

Berne, Emma Carlson.
  Hills / Emma Carlson Berne. — 1st ed.
      p. cm. — (Geography zone. Landforms)
  ISBN 978-1-4042-4207-4 (library binding)
  1. Mountains—Juvenile literature. I. Title.
  GB512.B475 2008
  551.43'6—dc22
                                                    2007034534

Manufactured in the United States of America

Contents

A hill is a large mound on Earth's **surface**. Hills are smaller than mountains. Hills can be made of dirt or they can be rock covered with dirt. Many hills have grass growing on the sides and the top. Generally, the top of a hill is rounded. Sometimes, it has a flat top. A flat-topped hill is sometimes called a butte or a mesa.

Hills are found on every **continent** in the world. Sometimes there are many hills near each other. Sometimes a hill will be by itself in the middle of flat land. Mountains often have hills near the bottom. These are called foothills.

The Painted Hills in Oregon are some of the most beautiful hills in the world. The hills spread out across 3,132 acres (1,267 ha).

Hills can be built in many ways. Earth's surface is made up of many huge pieces of rock called **plates**. These plates float on top of hot, melted rock. This means the plates are always moving around. Sometimes, they crash into each other. When the plates crash into each other, they can push the ground up where the plates meet. The pushed-up places are hills and mountains.

The movement of Earth's plates means that the surface of Earth is always changing. New hills will be formed over time and old ones will disappear.

These foothills of the Andes Mountains are in Chile. The Andes Mountains and their foothills were formed by the movement of Earth's plates.

Some hills are made by **glaciers**. Glaciers are huge sheets of ice that are always moving slowly across the ground. As the glaciers move, some ice melts into water. The water flows out from under the ice. It carries sand and rocks in it. The sand and rocks get piled up at the end of the glacier, making small hills.

Wind builds hills, too. The wind can blow dirt and sand around. Sometimes, the dirt and sand are dropped back onto the ground. When this happens in the same place over and over again, it makes hills.

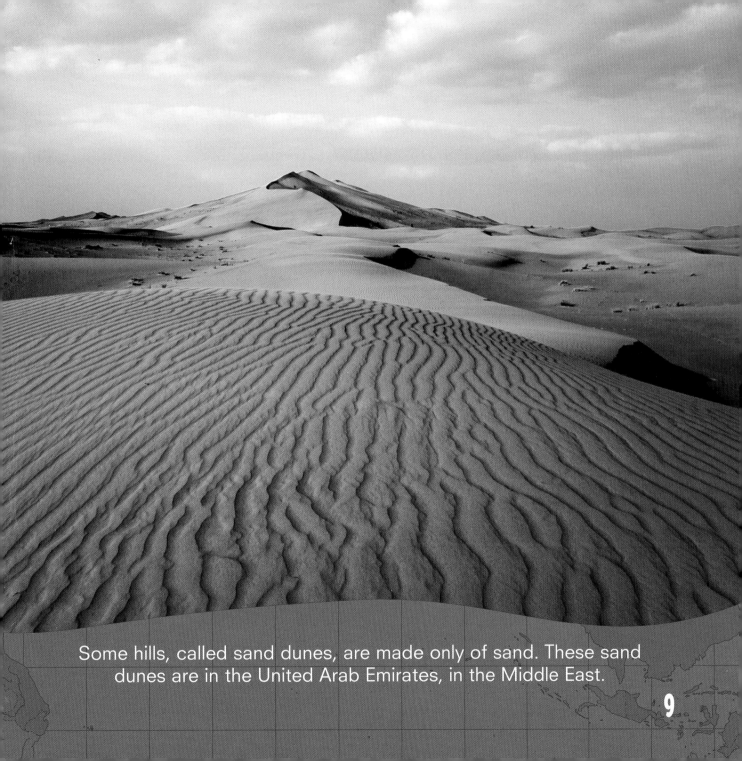

Some hills, called sand dunes, are made only of sand. These sand dunes are in the United Arab Emirates, in the Middle East.

Wind and rain can make hills through weathering and **erosion**. The wind can blow over giant mountains for millions of years. After a very long time, the wind wears away the high, sharp rocks of the mountains. The big mountains slowly turn into smaller, rounded hills.

Rain can turn mountains into hills also. Rain falling on mountains can slowly wear away the rock, just like the wind did. The rain washes away the bits of rock that it breaks away from the mountainsides. The mountains get smaller and smaller until they are hills.

This river and waterfall cut through the hills near Banos, Ecuador. Over time this water could make a canyon, or a deep, narrow valley, in the hills.

Volcanoes are special hills. They are made by lava. Lava is hot, melted rock.

In some places, there are **vents** in the surface of Earth. They are openings far down into Earth. Boiling hot, melted rock bubbles far under Earth's surface. Sometimes, the melted rock erupts, or shoots out of the vents as lava.

As the lava flows out of the vents it cools off. It gets hard and becomes rock. Over millions of years, the vent erupts many times. The lava keeps building up around the vent. After a long time, it makes a big hill of rock. This is the volcano.

This is Arenal volcano, in Costa Rica. One of the most active volcanoes in Costa Rica, Arenal grows taller every time it erupts.

Some hills are formed of loess. Loess is made by glaciers. When glaciers move, they sometimes crush rock into dust. This dust is called loess. Wind picks up the loess and carries it away in a cloud. When the wind dies down, the cloud of loess settles to Earth. It makes a mound. Over a long time, the wind leaves behind more dust. Eventually, the dust builds up into loess hills.

In Iowa, there are famous loess hills. They are 200 miles (322 km) long. They were made by glaciers and wind a long time ago, during the **Ice Age**.

These are the loess hills found in western Iowa and Missouri.
They have been eroded by wind and rain over time.

Hills provide a place for many plants and animals to live. Hills are often rocky and bumpy, though. Very large trees cannot grow on most hills. The soil is too thin and rocky. Smaller trees, shrubs, and grass often cover hills entirely, though.

Animals must be **nimble** to live on hills. Mountain goats like to climb up on the rocks on hillsides. Deer hide from **predators** among the trees. Mountain lions and bobcats like to live on high, steep hillsides, too. These large cats are very good at climbing and jumping.

The mountain goat uses its skill at climbing to find safe places
to eat and rest that other animals cannot reach.

The Black Hills in South Dakota are in the middle of flat plains. They are made of different kinds of rock, some hard and some soft. The Black Hills are also very old. Millions of years ago, the continental plates pushed the flat land up into mounds and folds.

In the middle of the Black Hills, a big block of **granite** sticks out. The wind and rain have eroded the soft rock around the granite. This block is called Mount Rushmore. The faces of four American presidents have been carved, or cut, into Mount Rushmore.

Mount Rushmore shows the faces of presidents George Washington, Thomas Jefferson, Theodore Roosevelt, and Abraham Lincoln.

The pine-covered Black Hills are 125 miles (201 km) long and 65 miles (105 km) wide. Deer, fish, wild sheep, coyotes, and mountain lions all live in the Black Hills. They share their home with many people who live and work there.

Many people work in the large coal and **mineral** mines in the Black Hills. Farmers raise sheep and cattle in the valleys between the hills. Some people also work cutting trees on the hills for **timber**.

Tourists also come to the Black Hills to visit. People like to walk, camp, hunt, and fish on the beautiful hills.

This prairie dog makes its home in the low hills and plains of South Dakota. Prairie dogs are kin to squirrels and live underground.

Hills are important to people all around the world. People often like to build villages and towns on hills. They can keep themselves safe from floods. Mosquitoes and other biting bugs that make people sick do not fly high up on hills. During wars, people often built forts on hills. This way, they could see their enemies coming from a long way off.

Hills are also very beautiful places. People love to visit hills, sing songs and write poetry about hills, and paint pictures of hills. All through history, people have loved the hills they live near.

Glossary

continent (KON-tuh-nent) One of Earth's seven large landmasses. The continents are North America, South America, Europe, Africa, Asia, Australia, and Antarctica.

erosion (ih-ROH-zhun) The wearing away of land over time.

glaciers (GLAY-shurz) Large masses of ice that move down a mountain or along a valley.

granite (GRA-nit) Melted rock that cooled and hardened beneath Earth's surface.

Ice Age (YS AYJ) A period of time about 12,000 years ago when it was much colder than it is today.

mineral (MIN-rul) A natural thing that is not an animal, a plant, or another living thing.

nimble (NIM-bul) Quick and not likely to fall when moving.

plates (PLAYTS) The moving pieces of Earth's crust.

predators (PREH-duh-terz) Animals that kill other animals for food.

surface (SER-fes) The outside of anything.

timber (TIM-bur) Wood that is cut and used for building houses, ships, and other wooden objects.

tourists (TUR-ists) People visiting a place where they do not live.

vents (VENTS) Openings that let air or other matter in or out.

Index

A
animals, 16

B
butte, 4

C
continent, 4

E
erosion, 10

F
foothills, 4

G
glaciers, 8, 14
granite, 18
grass, 4, 16

I
Ice Age, 14

M
mesa, 4
mound(s), 4, 14, 18
mountains, 4, 6, 10

P
plates, 4, 18

predators, 16

R
rock(s), 4, 6, 8, 10, 12, 14, 16, 18

S
surface, 4, 6, 12

T
timber, 20
tourists, 20

V
vents, 12

Web Sites

Due to the changing nature of Internet links, PowerKids Press has developed an online list of Web sites related to the subject of this book. This site is updated regularly. Please use this link to access the list:
www.powerkidslinks.com/gzone/hill/